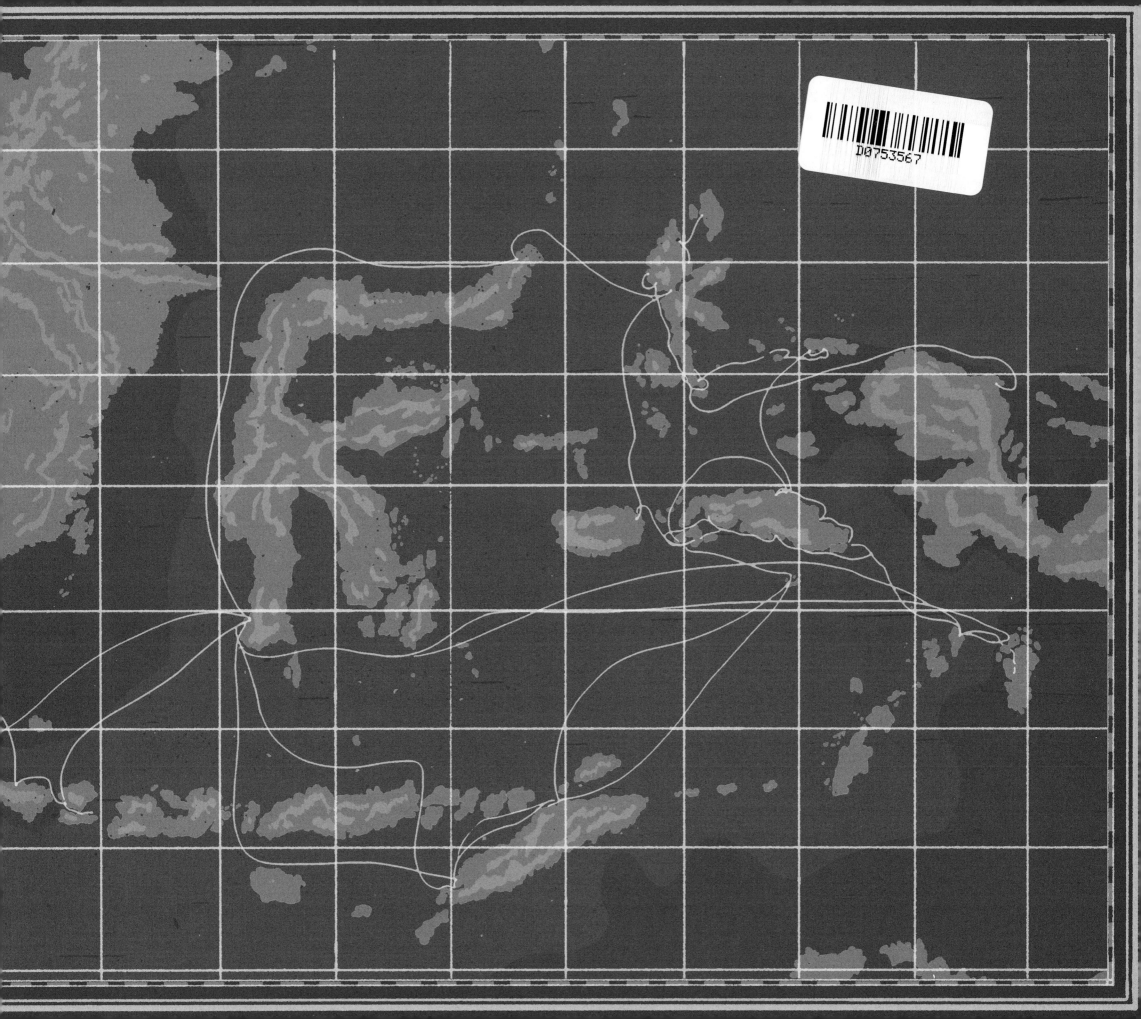

DARWIN'S RIVAL
ALFRED RUSSEL WALLACE
AND THE SEARCH FOR EVOLUTION

CHRISTIANE DORION

ILLUSTRATED BY
HARRY TENNANT

For Philip, Thomas, and Nicholas
C. D.

Dedicated to the memory of Finn Clark
H. T.

Historical consultant: Dr. George Beccaloni, director of the Alfred Russel Wallace Correspondence Project.
Thank you to Martin Hinchcliffe for your passion and expertise on Victorian scientific instruments.

Text copyright © 2020 by Christiane Dorion
Illustrations copyright © 2020 by Harry Tennant

First U.S. edition 2020

Library of Congress Catalog Card Number pending
ISBN 978-1-5362-0932-7

19 20 21 22 23 24 LEO 10 9 8 7 6 5 4 3 2 1

Printed in Heshan, Guangdong, China

This book was typeset in Gill, Nevis, and WB Harry Tennant.
The illustrations were created digitally.

Candlewick Studio
an imprint of
Candlewick Press
99 Dover Street
Somerville, Massachusetts 02144

www.candlewickstudio.com

CANDLEWICK STUDIO
an imprint of Candlewick Press

Charles Darwin

Downe House

Lusted Road

D...

In June 1858, the prominent scientist Charles Darwin received a letter from a young British explorer and naturalist. The letter had been sent from a remote island in the Pacific Ocean, and it proposed a new idea to explain how living things evolved over time. The explorer's name was Alfred Russel Wallace, and this is the story of his life, his adventures, and how his letter to Darwin had a huge impact on our understanding of life on Earth.

Charles Darwin

THE YOUNG ALFRED

ALFRED RUSSEL WALLACE was born on January 8, 1823, near the village of Llanbadoc in England, now part of Wales. He was the second youngest of nine children, and his family lived in a cottage on the banks of the River Usk. It was the dawn of the Industrial Revolution, a time of great discoveries, inventions, and ideas. The dark smoke of factory chimneys dotted the British landscape, and coal mines were dug deep beneath the ground to meet the growing demand for this important source of power. Towns were growing into industrial cities. The introduction of the first steam locomotive and the use of iron in shipbuilding were making travel and trade around the world faster and easier.

Born in the countryside, Alfred was sheltered from these rapid changes to life in Britain. As a young boy, he would spend hours climbing the steep wooded slope behind the family home and exploring the surrounding fields with his older brothers and sisters. He also loved playing in the river, catching small fish with an old saucepan, and roasting potatoes over a campfire.

This all changed when Alfred was five years old. Due to financial difficulties, the family moved to Hertford, a small town north of London. There, his father took a job as a librarian. He would also tutor pupils to supplement his modest income. Their new home was a much smaller townhouse with a tiny garden, but Alfred enjoyed having close neighbors to play with for the first time. There was also a constant supply of interesting books at home, which his father would read aloud to the family in the evening. Alfred went to a local school, with eighty boys and four teachers all crammed into a single

classroom. Although he had an inquisitive mind, he found memorizing times tables, Latin verbs, and the names of kings and queens rather tedious. He much preferred reading travelers' tales and stories of shipwrecks on desert islands.

Money was always short and the family had to move from one house to another, each time living in smaller and more cramped conditions. Alfred was only fourteen when he had to leave school to earn a living. But his father's financial troubles and the constant lack of money helped to shape his character and made him all the more determined to make his own way in the world.

THE MAKING OF AN EXPLORER

IN THE SUMMER OF 1837, Alfred began to learn a trade with his eldest brother, William, who was a land surveyor. They worked together for more than six years, traveling across southern England and Wales making detailed maps for landowners and the expanding railway. Each day, they would set out with their sextant, measuring rod, chain, and flags to map the boundaries of fields, woods, roads and streams, and the positions of buildings. This solitary work suited Alfred's nature. He enjoyed working in the countryside, tramping the hills and breathing the fresh air. His earnings were just enough to pay his bills, and he learned skills that would come in handy a few years later.

Alfred began to develop a keen interest in nature and science. He wanted to know more about the various flowers, shrubs, and trees he saw every day, so he started collecting plants. After a hard day's work, he would spend his evenings drying and identifying his most precious findings. His brother William thought it was all a waste of time, but this would turn out to be another useful skill in the future.

When he was twenty, a shortage of surveying work forced young Alfred to leave his brother and take a teaching job in a boys' boarding school in Leicester. Alfred was extremely shy, so it was a rather daunting experience, and he soon decided that teaching wasn't for him. However, it was during this time that he met Henry Walter Bates, who shared his passion for nature, introduced him to the joy of collecting beetles, and would become his lifelong friend. Full of enthusiasm, Alfred bought himself a book on British insects, as well as bottles, pins, and a wooden box to store his specimens, and he began collecting avidly. He was fascinated by the different shapes, range of colors, and huge variety of beetles that could be found locally. "Why do they all look so different?" he wondered. "Where did they all come from?"

> "I begin to feel rather dissatisfied with a mere local collection—little is to be learnt by it. I should like to take some one family (of insects), to study thoroughly—principally with a view to the theory of the origin of species."

Teaching did give Alfred time to visit the local library and read everything he could find on natural history. In those days, most people believed that all species of plants and animals on Earth had been created in their current form by divine intervention. However, some naturalists were now proposing the daring idea that living things gradually changed or evolved over time. How this happened and why remained a mystery that captivated Alfred. He also enjoyed reading about travelers' adventures to foreign places, and he longed to see the wonders of the tropical world. He was particularly intrigued by a British naturalist named Charles Darwin, who had just returned from a long voyage around the world. Darwin had collected an incredible variety of plants, animals, rocks, and fossils on his five-year journey aboard HMS *Beagle* and was busy writing about his discoveries. His journal captured Alfred's imagination with its rich descriptions of amazing animals, from extinct giant sloths to marine iguanas and duck-billed platypuses.

Before long, Alfred was planning his own expedition with his friend Henry Bates to a part of the world that was largely unknown to Europeans—the Amazon rain forest. Alfred was twenty-five and Henry was twenty-three. Unlike Darwin and most adventurers at the time, the two young men didn't have family wealth or friends in high places to fund their expedition. They had to find a different way. With the spread of exploration and trade, Victorians had a fascination for foreign wildlife and displayed collections of insects and birds as a popular pastime. Collectors would pay good prices for real specimens, even for the tiniest bugs. Alfred's plan was to make a living by finding rare and previously unrecorded species of beetles, butterflies, and birds and selling them to museums and keen amateurs in England. But mostly, he was hoping to gather his own collection and find clues that would help him understand how species changed over time.

Neath, Wales, where Alfred and William worked as surveyors

Liverpool,
Great Britain

NORTH
AMERICA

MAY 28, 1848

Belém,
Brazil

SOUTH
AMERICA

ABOARD THE *MISCHIEF*

WALLACE AND BATES met up in London to prepare for their long voyage. They bought the necessary equipment and visited the British Museum to sketch and make notes on the more valuable species of insects and birds. They were assured by the museum that the Amazon rain forest would provide many rare species and that they would be able to cover their expenses by selling what they collected. They also found an agent, Samuel Stevens, who would arrange for the sale of specimens and help them publish their findings in scientific magazines. The friends had just enough money to pay for their passage on board a small cargo ship, the *Mischief*.

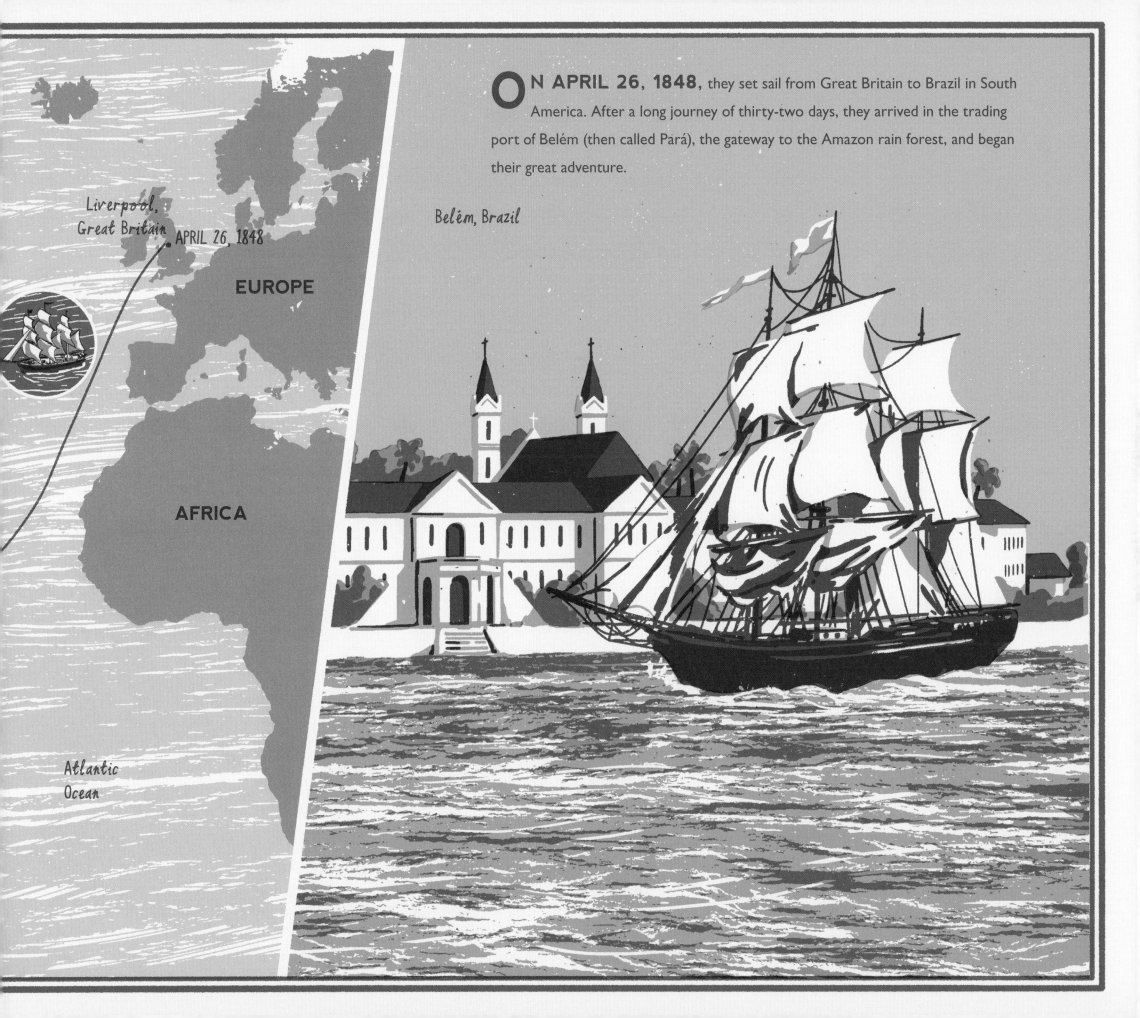

ON **APRIL 26, 1848,** they set sail from Great Britain to Brazil in South America. After a long journey of thirty-two days, they arrived in the trading port of Belém (then called Pará), the gateway to the Amazon rain forest, and began their great adventure.

Liverpool, Great Britain
APRIL 26, 1848

EUROPE

AFRICA

Atlantic Ocean

Belém, Brazil

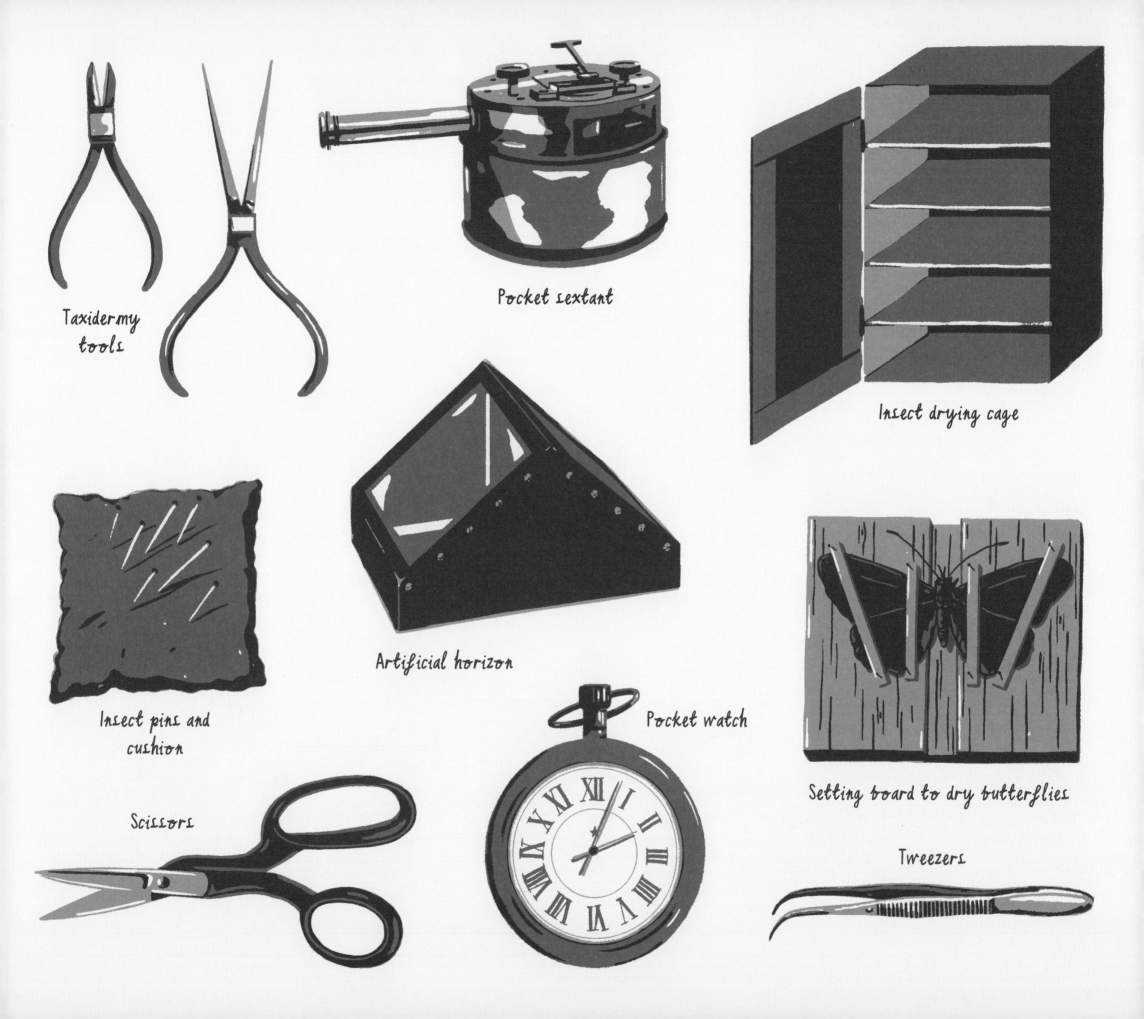

Taxidermy tools

Pocket sextant

Insect drying cage

Insect pins and cushion

Artificial horizon

Setting board to dry butterflies

Scissors

Pocket watch

Tweezers

Collecting net

Magnifying lens

COLLECTORS' EQUIPMENT

Barometer

Stoneware jars for preserving reptiles and fish in alcohol

Small collecting box lined with cork

Birdcage

Thermometer

Prismatic compass

Lantern

FIRST STEPS IN THE RAIN FOREST

THE YOUNG EXPLORERS found a comfortable house to rent outside Belém, with a large veranda where they could sit and work. They hired a cook and bought a table, a few chairs, and some hammocks. Armed with their nets, bottles, and collecting boxes, they started exploring the surrounding rain forest in search of rare beetles, dazzling butterflies, and unusual birds. They were hoping to discover many species that had never been seen before in Europe.

The rain forest was magnificent. There was a rich variety of palm trees, giant vines hanging from branches, and creepers with large, shiny leaves climbing toward the sun. However, Wallace was slightly disappointed at first. He was expecting to see a profusion of brightly colored parrots and monkeys leaping from branch to branch, as had been so vividly described by previous travelers. But he learned to listen to the sounds of the forest and look carefully for creatures hidden among the gigantic tree roots and leafy canopy.

The collectors quickly fell into a rigorous daily routine. Up at dawn, they would spend the first two hours looking for birds, enjoying the clear sky and cooler temperature of the early morning. Then they would have breakfast and search for beetles and butterflies from ten until two. Every insect caught was pinned and put away in a collecting box. In the afternoon, the daily tropical showers forced them to rest for a few hours. A late lunch was served at four, followed by a light meal at seven. The evenings were spent preparing and labeling all their specimens. Soon the veranda began to fill up with a collection of live creatures, including a large and hairy bird-eating spider, a sleepy sloth, and a boa constrictor that, according to Wallace, sounded rather like "high-pressure steam escaping from a Great Western locomotive."

Wallace was excited by the things he discovered every day and filled his journal with careful observations and detailed sketches. He often compared the loud chorus of the rain forest to the sound of factories and steam engines back home. He especially enjoyed the concert of frogs taking over from the singing of cicadas and the howling of monkeys at twilight. "There are three kinds, which can frequently be all heard at once," he noted. "One of these makes a noise . . . one would expect a frog to make, namely a dismal croak, but the sounds uttered by the others were like no animal noise. . . . A distant railway train approaching, and a blacksmith hammering on his anvil, are what they exactly resemble." He also likened the early morning calls of the parakeets echoing from the trees to "a hundred knife-grinders at full work."

> "In all works on natural history, we constantly find details of the marvellous adaptation of animals to their food, their habits, and the localities in which they are found. But naturalists are now beginning to look beyond this, and to see that there must be some other principle regulating the infinitely varied forms of animal life."

Once they had explored the area around Belém, Wallace and Bates began to venture farther afield on longer expeditions, sailing up and down the nearby rivers and staying in villages and on plantations and cattle ranches along the way. Their collection grew quickly and, as they hoped, included many insects new to European science. After two months, they sent the first shipment to their agent back in England—more than a thousand different species of beetles and butterflies, all carefully pinned and labeled inside large wooden boxes. These would be sold to museums and collectors to enable Wallace and Bates to continue their expedition.

Wallace's house
near Belém

"Here no one who has any feeling of the magnificent and the sublime can be disappointed; the sombre shade, scarce illumined by a single direct ray even of the tropical sun, the enormous size and height of the trees, most of which rise like huge columns a hundred feet or more without throwing out a single branch, the strange buttresses around the base of some, the spiny or furrowed stems of others, the curious and even extraordinary creepers and climbers which wind around them . . . altogether surpass description, and produce feelings in the beholder of admiration and awe."

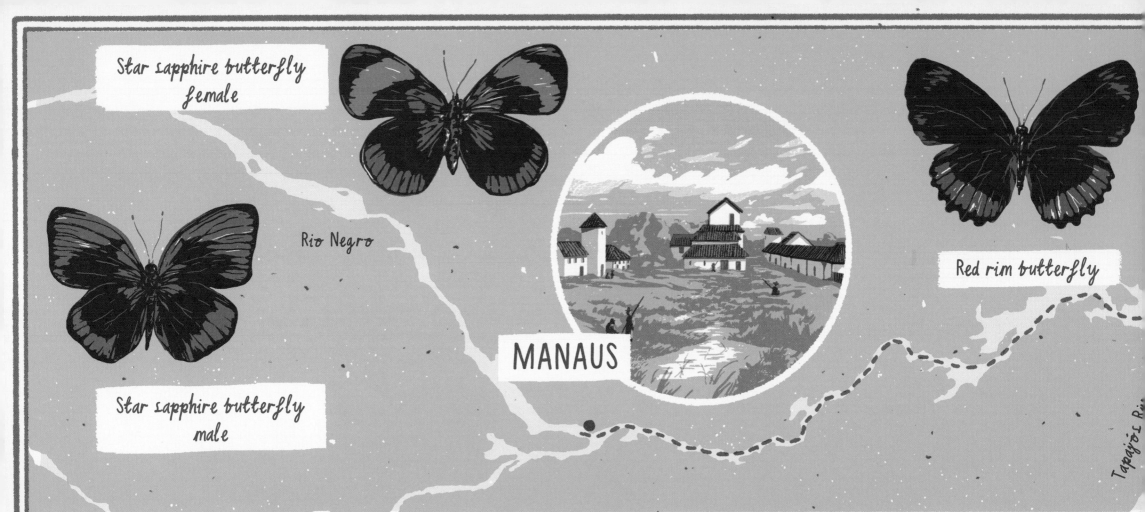

Star sapphire butterfly female

Rio Negro

Red rim butterfly

MANAUS

Star sapphire butterfly male

Tapajós River

SAILING ON THE AMAZON

AFTER A FEW MONTHS exploring together, Alfred and Henry split up to collect in different areas. Alfred's plan was to sail up the Amazon and then into the farthest reaches of the Rio Negro, where few Europeans had ventured. Tired of the humdrum of factory work back home, his younger brother, Herbert, came to join him for this new expedition.

In August 1849, the brothers found passage on a small trading boat going to the town of Santarém on the Amazon River. The boat had a leaky deck and a strong smell of fish and animal skins, but "voyagers on the Amazon must not be fastidious," commented Alfred. Battling strong currents, they sailed slowly through the maze of islands and narrow channels at the entrance of the river. When the wind fell, the crew had to paddle hard against the current and even pull the boat with a rope from the shore.

After twenty-eight days, they reached Santarém, perched on a slope above a sandy beach, with its whitewashed houses and red tiled roofs. It was the dry season, with very little rain and a clear blue sky, and Alfred was eager to explore the area. The brothers made their base in a small mud hut and started collecting, sometimes following the banks of the river, sometimes delving deeper into the forest. Butterflies were abundant but difficult to catch, settling high up in the trees. Each capture was a thrill for Alfred, and he was enchanted by the stunning colors of these exquisite insects, describing them as "spangled with gold, or glittering with the most splendid metallic

Amazon River

Atlantic Ocean

Postman butterfly

BELÉM

SANTARÉM

Tocantins River

tints." The rarest and prettiest would fetch the highest prices among collectors. Three more boxes of butterflies and other specimens, including a small alligator, were shipped to England to sell for further funds.

Alfred was very meticulous in the way he labeled his specimens, always recording where they were found. Collecting on both banks of the Amazon, he started to notice that the wide river formed a barrier, with different species of monkeys on either side. The same was true for some species of birds and butterflies. It was only years later, while collecting in another part of the world, that the significance of this observation would become clear.

In early November, the constant rain made it harder to collect

"Places not more than fifty or a hundred miles apart often have species of insects and birds at the one, which are not found at the other. There must be some boundary which determines the range of each species; some external peculiarity to mark the line which each one does not pass."

in the forest, so the brothers prepared to continue their journey up the Amazon toward the Rio Negro. They gathered provisions and repaired their leaky canoe. They also recruited local men to help them navigate the river and carry their equipment. Despite the downpours and swarms of tormenting mosquitoes, they made slow but steady progress, staying in villages along the way and paddling against the current more often than sailing. They finally reached Manaus (then called Barra), at the junction of the Amazon and Rio Negro, on December 31, 1849.

EXPLORING THE RIO NEGRO

ALFRED AND HERBERT stayed around Manaus for eight months, but their collection of insects and birds disappointed them. The only specimen they thought was valuable was the rare and peculiar umbrellabird, found on the islands of the river. It was as black as a raven, with a long wattle on its chest and a crest of feathers covering its head like an umbrella.

At the end of the summer, after receiving more funds from England, Alfred decided to continue his journey up the Rio Negro, while his brother chose to slowly make his way back home. Herbert felt that he was not cut out for the hard life of collecting. He was more interested in writing poetry than searching for butterflies and birds. Sadly, Alfred would discover several months later that his younger brother had not made it home but had died of yellow fever, a tropical disease spread by mosquitoes.

On August 31, 1850, Alfred left Manaus on the boat of a local trader heading upriver, Senhor João Antonio de Lima. The boat was crammed with provisions and trading goods for the local population, including fishhooks, buttons, needles, thread, axes, and bales of cotton. It was a large canoe with a rough deck made of palm trees and a curved thatched roof, with just enough room to sit or lie down comfortably among the many wooden cases. After over two years in Brazil, Wallace was ready to explore the part of the Amazon rain forest previously unknown to Europeans.

The water of the Rio Negro was inky black and looked very different from the yellow, muddy Amazon. The river was dotted with small rocky islands and bordered by rugged shores and an unbroken forest. But most of all, it was mosquito-free, to Wallace's great delight!

After a while, the landscape began to change, with a few rocky peaks rising above the thick forest. A series of rapids, falls, and dangerous whirlpools made navigation on the river difficult, and the boat had to be swapped for smaller canoes. The men had to paddle furiously against the strong current, zigzagging from bank to bank, often having to unload the canoes and pull them by hand through narrow channels. Where many would have felt defeated, Alfred Wallace remained full of hope and enthusiasm. "The brilliant sun, the sparkling waters, the strange fantastic rocks, and broken woody islands were a constant source of interest and enjoyment," he wrote.

The rare umbrellabird

> "There are immense whirlpools which engulf large canoes. The waters roll like ocean waves, and leap up at intervals, forty or fifty feet into the air, as if great subaqueous explosions were taking place."

"Three of these (rapids) were very bad, the canoe having to be unloaded entirely, and pulled over the dry and uneven rocks. The last was the highest; the river rushing furiously about twenty feet down a rugged slope of rock."

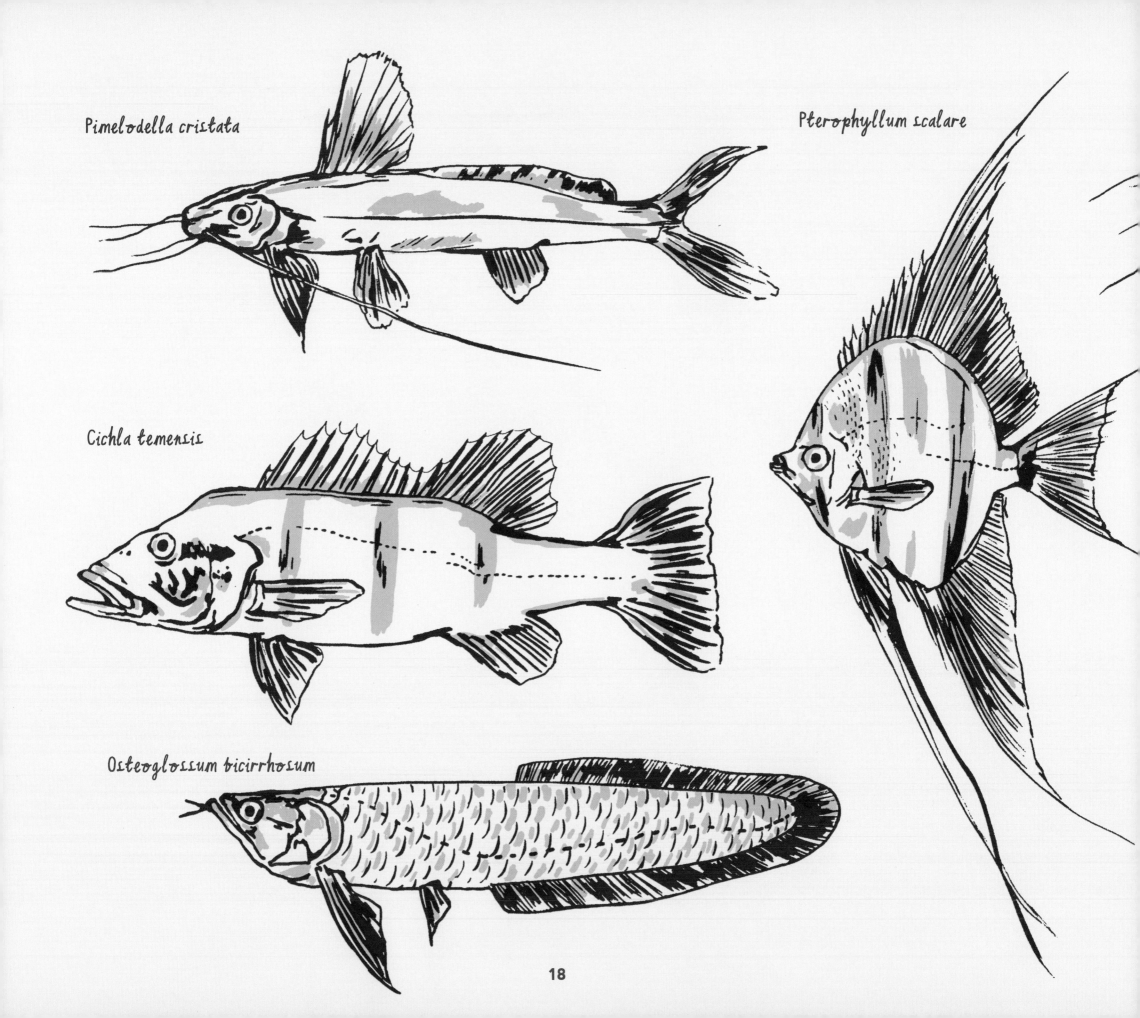

Pimelodella cristata

Pterophyllum scalare

Cichla temensis

Osteoglossum bicirrhosum

18

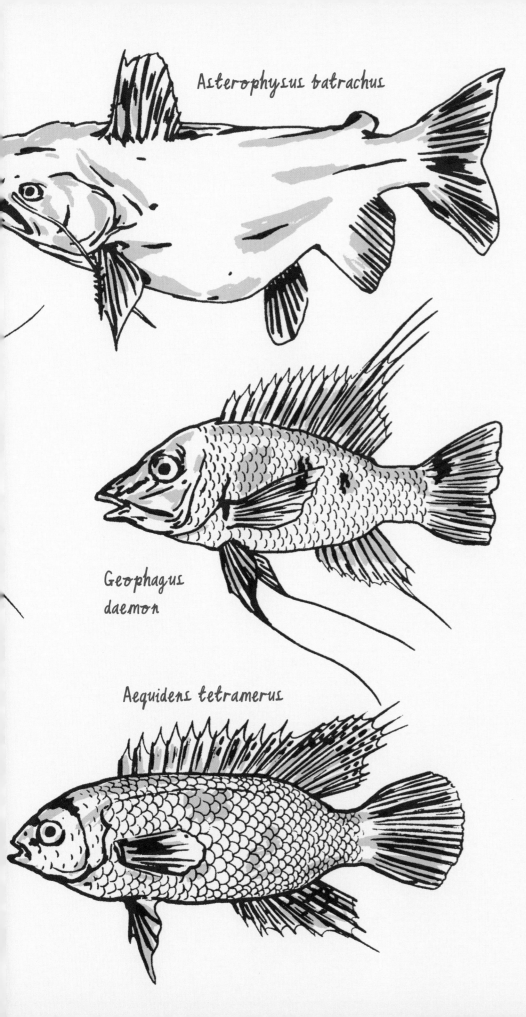

Asterophysus batrachus

Geophagus
daemon

Aequidens tetramerus

Being on the water for so long, Wallace turned his interest to the huge variety of fish caught in the river. Their curious shapes and patterns captivated him—some had long, sharp spines, others had curious markings or huge fins. Some inflated like a ball when facing danger. Wallace was fascinated by how the fish differed from one river to another. He was starting to see patterns in the distribution of species without being able to explain why.

"Of all kinds of fishes I found two hundred and five species in the Rio Negro alone, and these, I am sure, are but a small portion of what exist there. Being a black-water river, most of its fishes are different from those found in the Amazon. In fact, in every small river, and in different parts of the same river, distinct kinds are found."

Two months later, on October 24, 1850, they reached the small village of Nossa Senhora da Guia, where Senhor Lima lived. Wallace was invited to stay for as long as he wished and was loaned a canoe to explore the area. Soon he was setting off on short expeditions with a few local men, following streams into the forest and climbing steep rocky peaks. With the help of the local people, Wallace encountered a magnificent bird whose plumage he described as "a mass of brilliant flame." This was the rare Guianan cock-of-the-rock, coming to breed in the mountain caves. He was thrilled to be able to observe the behavior of these unusual birds in the wild, the males taking turns to dance and display their extraordinary feathers to impress the females.

At the end of January, the wet season had set in and Wallace decided to continue his expedition. This time he took very little equipment with him: his butterfly net, a collecting box, and a few scientific instruments as well as goods that could be exchanged for food or rare specimens. The journey was incredibly difficult, with dangerous rapids and a leaky canoe. There were also stinging ants, biting flies, and bloodsucking bats to endure. Despite all this, Wallace enjoyed staying with local people deep in the forest and learning more about their way of life. He was fascinated by their peaceful communities and ability to live in harmony with their surroundings. Wallace thought he must have been a source of curiosity because he did not look like the people who lived in the Amazon rain forest. "A hundred bright pairs of eyes were continually directed on me from all sides, and I was doubtless the great subject of conversation," he wrote. Wallace's sense of adventure extended to adding local dishes to his regular diet of fish and fruit. He found agouti, a relative of the wild guinea pig, rather dry and tasteless, but enjoyed fried monkey, which he described as tasting "a little bit like rabbit." An alligator tail, turtle stew, and roasted anaconda also made very acceptable meals.

The butterflies were prolific, and Wallace was astonished by the rich variety of life in the forest. He came across animals he had never seen before, including many fine tropical birds and a few poisonous snakes. He also came face-to-face with an animal he had long wished to meet—a fine black jaguar. Rather than being scared by this encounter, Alfred Wallace felt only admiration for such an impressive creature. "In the middle of the road he turned his head, and for an instant paused and gazed at me," wrote Wallace, "but having, I suppose, other business of his own to attend to, walked steadily on, and disappeared in the thicket."

While navigating the Rio Negro, Wallace produced with great accuracy the first detailed map of the mighty river, using the skills he had learned as a surveyor and a few simple scientific instruments. With a compass and sextant he was able to pinpoint different positions on the river. Unable to see the horizon in the dense forest, he used an instrument called an artificial horizon that measured the angle of the sun and stars in the sky from their reflection in a small tray of liquid. With his watch, he was able to work out distances by measuring the time for a canoe to travel from one point to another. Local people also provided him with invaluable information, as they knew every bend and twist of their river.

By March 1852, Wallace was suffering from constant attacks of fever, probably caused by malaria. Having been away for four years, he was now longing to see his family and the green fields of England. He began the long journey back down the river with his many boxes of specimens and a rather large collection of live monkeys, parrots, macaws, and other very noisy animals. He reached Belém on July 2, 1852, ready to sail back home.

> "It is here that the rarest birds, the most lovely insects, and the most interesting mammals and reptiles are to be found. Here lurk the jaguar and the boa constrictor, and here amid the densest shade the bell-bird tolls his peal."

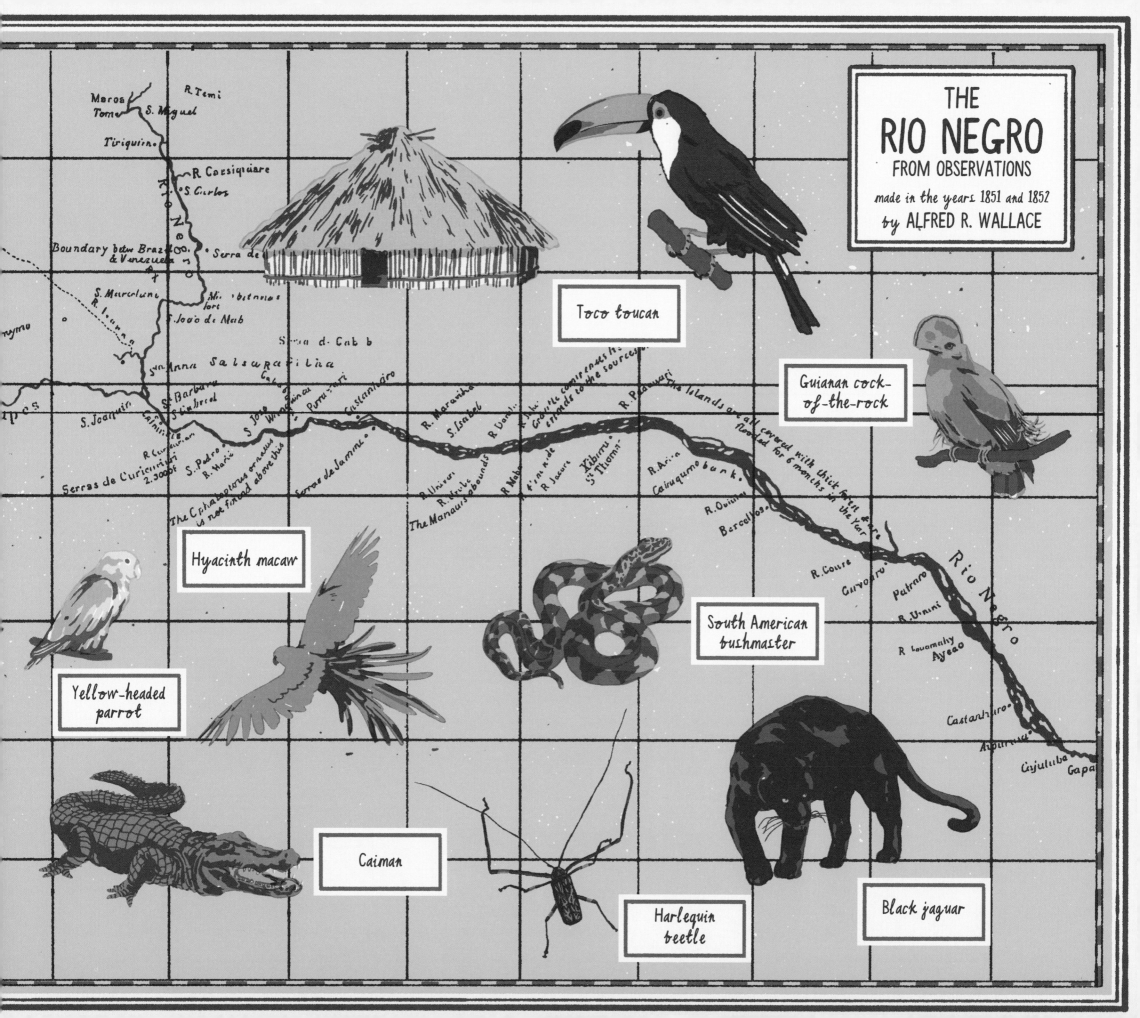

THE RIO NEGRO
FROM OBSERVATIONS
made in the years 1851 and 1852
by ALFRED R. WALLACE

Toco toucan

Guianan cock-of-the-rock

Hyacinth macaw

Yellow-headed parrot

South American bushmaster

Caiman

Harlequin beetle

Black jaguar

SHIPWRECKED

ON JULY 12, 1852, Alfred Wallace waved goodbye to the white houses and swaying palm trees of Belém and set sail on board the *Helen* with his collection of specimens and live animals. The ship was carrying a large cargo of rubber, cocoa, palm wood, and balsam, a tree resin used in varnish. Apart from the crew, Wallace was the only passenger on board, and he was too ill to leave his cabin, suffering from fever and severe seasickness with every roll of the ship. Twenty-six days into the journey, in the middle of the Atlantic, disaster struck! It was just after breakfast when the captain burst into Wallace's cabin to announce that the ship was on fire. The highly flammable balsam cargo had been poorly stored and had caught fire. After a relentless effort to put out the flames, the captain gave the order to abandon ship. Wallace struggled back into his smoky cabin to rescue a few things—his watch and a tin box containing his map of the Rio Negro, some drawings, a couple of shirts, and the little money he had.

The crew lowered two lifeboats into the water and filled them with provisions, including bread, preserved meat, biscuits, and a few barrels of water. Wallace, still feeling weak and ill, slipped down a rope and stumbled into one of the boats. The lifeboats were leaky, having been exposed to the tropical sun for so long, and the passengers had to bail water out constantly with buckets. They stayed near the burning ship all night, hoping that the fire would attract a passing boat, and Wallace was forced to watch his precious collection of thousands of insects and birds go up in flames in "a fiery furnace tossing restlessly upon the ocean." Typical of his character, he remained calm, putting out of his mind what he had lost. After all, everyone had been saved. Following a long night, they

set sail toward the nearest land, more than a thousand kilometers away.

The castaways traveled for days, blistered by the scorching sun and soaked by constant spray. The waves were high and strong gusts of wind frequently heeled the lifeboats over dangerously. Even in these desperate times, Wallace showed an endless sense of wonder for the natural world, admiring flying fish jumping above the waves, dolphins with their beautiful metallic shades, and seabirds circling above with their broad wings. He also loved watching shooting stars in the night sky. What better place for observing them "than lying on my back in a small boat in the middle of the Atlantic," he would recall.

After ten long days, they were finally rescued by an old cargo ship, the *Jordeson*,

making its way from Cuba to England and overloaded with heavy timber. Unfortunately, Wallace's troubles were not over yet! The journey was slow, provisions were short, and they were running out of drinking water. Approaching the coast of England, the ship was caught in a violent storm and nearly sank, with water pouring over its rotten decks. The situation was so precarious that the captain slept with an ax by his bed in case a mast needed to be cut down to avoid capsizing. At last, after an extremely difficult voyage of about eighty days, they arrived in Deal, Kent, on October 1, 1852. With nothing left but the clothes on his back and the small tin box rescued from the fire, Wallace vowed that he would never venture to sea again. But within days of being back home, he was already thinking about his next expedition.

> "Everything was gone, and I had not one specimen to illustrate the unknown lands I had trod, or to call back the recollection of the wild scenes I had beheld!"

"It was now a magnificent spectacle, for the decks had completely burnt away, and as it (the ship) heaved and rolled with the swell of the sea, presented its interior towards us filled with liquid flame — a fiery furnace tossing restlessly upon the ocean."

BACK IN ENGLAND

ON HIS ARRIVAL, Wallace was met by his agent, Samuel Stevens. The first thing Wallace did was buy some clothes and have a new suit made by a tailor. Luckily, Stevens had insured the lost collection, and there was also some money left over from the sale of specimens already sent home. Wallace rented a house near Regent's Park, a few steps from the London Zoo, and settled down comfortably in his new home.

There was growing curiosity in Britain about natural history, and people were eager to learn more about foreign wildlife and faraway places. Wallace set out to write his first book on palm trees, using the detailed sketches he had recovered from the shipwreck. A second book about his travels and observations in the Amazon soon followed.

Wallace's achievements were becoming recognized, and he was often invited to meet other scientists. This helped him gain support for his next expedition. He even had the pleasure of briefly meeting

CHARLES DARWIN English naturalist who explored some of the most remote places on Earth during his five-year voyage aboard the Beagle, and gathered a huge collection of plants, animals, rocks, and fossils.

CHARLES LYELL Scottish geologist who was one of the first to propose that rocks form and wear away due to natural forces, gradually changing Earth's surface over long periods of time.

for the first time a man whose travels had greatly inspired him: Charles Darwin. At the time, Darwin was forty-four years old and living comfortably off family wealth in a large country house with his wife and children. He was now a well-established naturalist, busy studying the fossils and specimens he had collected during his voyage on the *Beagle* and writing about his findings.

After just over a year in England, Wallace was planning a new expedition to the Malay Archipelago, the thousands of islands between the Indian and Pacific Oceans. This was one of the wildest and least explored parts of the world, known for its rich wildlife. So that he knew what to look for, Wallace studied existing collections of birds and insects at the British Museum and made notes of the rarest and most valuable specimens. This time, after persistent requests to the Royal Geographical Society, Wallace was offered a first-class ticket on modern paddle steamers to Singapore. Gone were the days of traveling on old, leaky cargo ships! He also hired a young assistant named Charles Allen to accompany him on his journey and help him with his work.

JOSEPH HOOKER English botanist who traveled from the Antarctic to the Himalayas to collect plants, many new to European scientists. Later on in his life he became director of the Royal Botanic Gardens in Kew, London.

THOMAS HUXLEY English naturalist who developed his interest in sea anemones, jellyfish, and other marine invertebrates while traveling as an assistant surgeon on the Royal Navy ship HMS *Rattlesnake*.

A NEW EXPEDITION

ON MARCH 4, 1854, Alfred Wallace
and Charles Allen left Portsmouth on board
the *Euxine,* headed for Singapore. The voyage took
forty-five days and included a short journey by
barge down the Nile River, crossing the desert
in horse-drawn carriages, and two changes
of boat. Wallace was thrilled to see so many
exciting places on the way, especially the tombs
of the Knights of Malta, the pyramids of Egypt,
and the vast desert with its camel trains. After a
fascinating and comfortable voyage, they landed
on the island of Singapore on April 18, 1854.

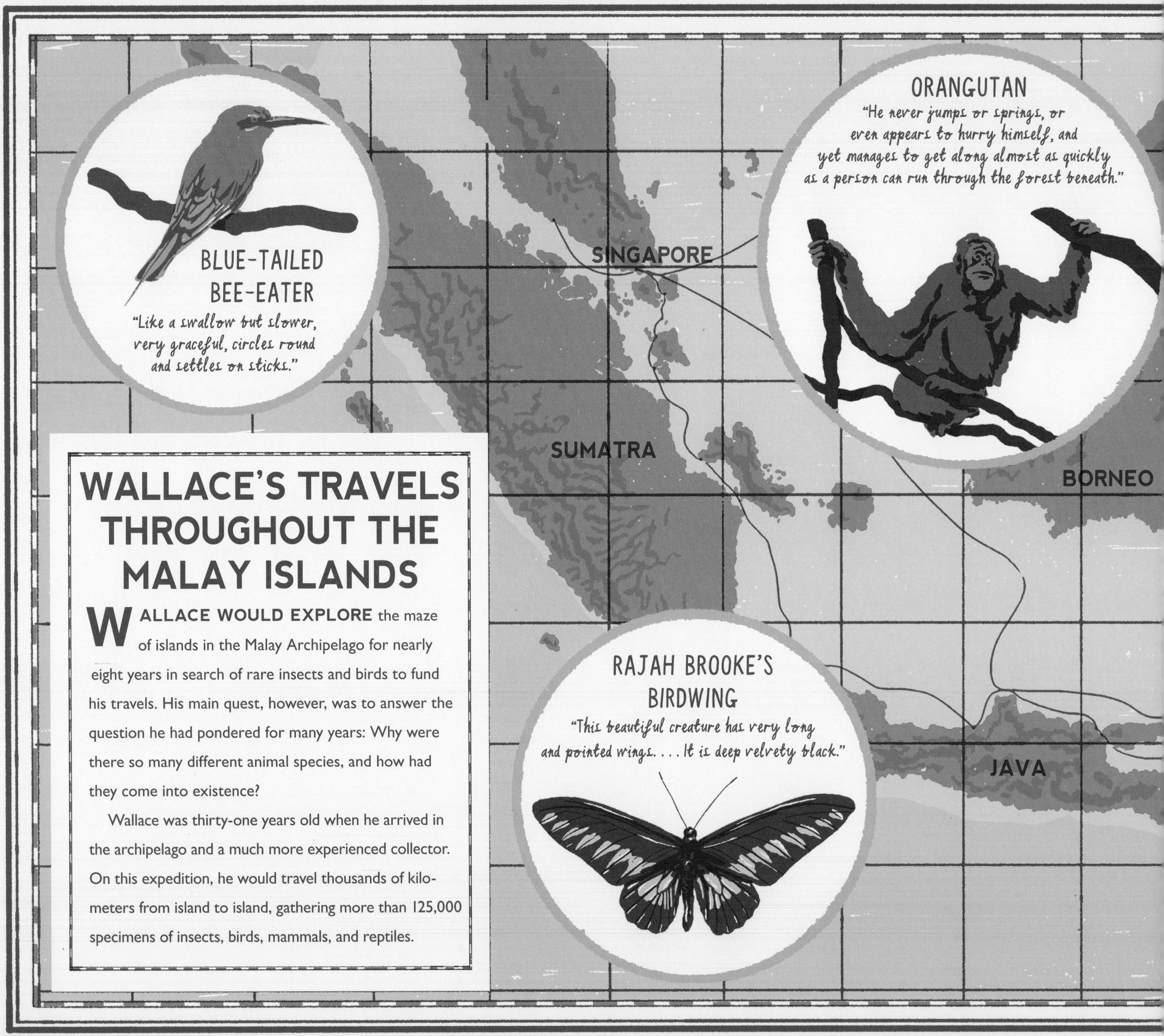

BLUE-TAILED
BEE-EATER

"Like a swallow but slower,
very graceful, circles round
and settles on sticks."

ORANGUTAN

"He never jumps or springs, or
even appears to hurry himself, and
yet manages to get along almost as quickly
as a person can run through the forest beneath."

SINGAPORE

SUMATRA

BORNEO

WALLACE'S TRAVELS THROUGHOUT THE MALAY ISLANDS

WALLACE WOULD EXPLORE the maze of islands in the Malay Archipelago for nearly eight years in search of rare insects and birds to fund his travels. His main quest, however, was to answer the question he had pondered for many years: Why were there so many different animal species, and how had they come into existence?

Wallace was thirty-one years old when he arrived in the archipelago and a much more experienced collector. On this expedition, he would travel thousands of kilometers from island to island, gathering more than 125,000 specimens of insects, birds, mammals, and reptiles.

RAJAH BROOKE'S
BIRDWING

"This beautiful creature has very long
and pointed wings. . . . It is deep velvety black."

JAVA

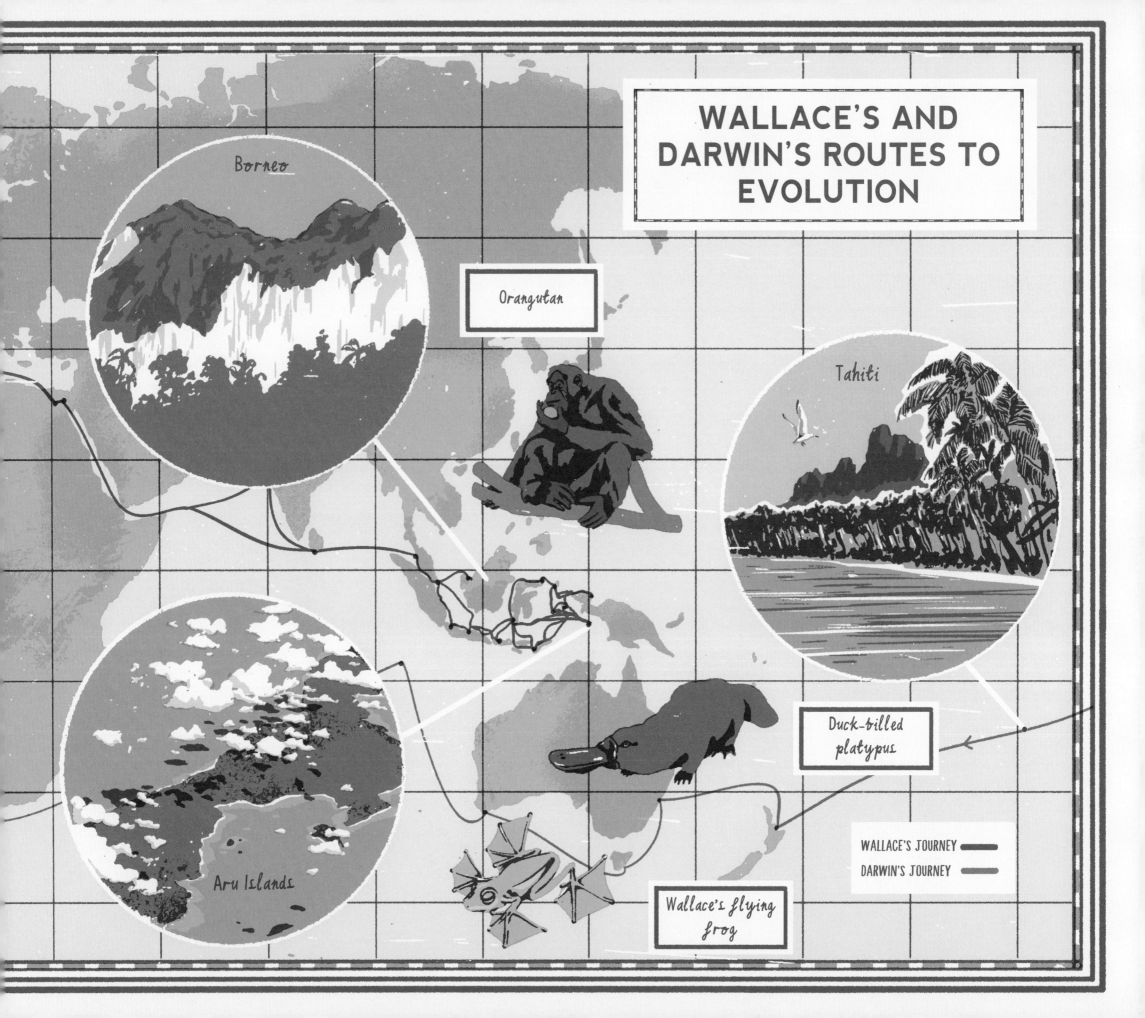

WALLACE'S AND DARWIN'S ROUTES TO EVOLUTION

Borneo

Orangutan

Tahiti

Aru Islands

Duck-billed platypus

Wallace's flying frog

WALLACE'S JOURNEY
DARWIN'S JOURNEY

BACK HOME

WALLACE HAD TRAVELED for nearly eight years across the Malay Archipelago and, at the age of thirty-nine, he was now ready to return home. Waiting in Singapore for a passage to England, he bought two handsome birds of paradise in a large bamboo cage for the London Zoo. These birds had rarely been seen alive in Europe and would provide him with a much-needed income. Caring for his precious cargo on the long passage home was a challenge; Wallace had to feed the birds with fruit and cockroaches he caught on the ship. After a long voyage, Wallace arrived in London on March 31, 1862.

Wallace moved in with his sister Fanny and her husband and slowly settled back into the British way of life. He recovered the many boxes of specimens he had sent home during his expedition for his own collection and was thrilled to rediscover the thousands of insects and birds he had caught in the rain forest. The first few months were spent arranging and studying his specimens, as well as making frequent visits to the British Museum, where many of his specimens were already in the museum's collection. He also spent a great deal of time catching up with his family and sorting out his own affairs.

Wallace was surprised and delighted to receive such a warm welcome from the scientific community; he was now seen not only as a brilliant collector, but also as a great naturalist in his own right. Over the years, he became close friends with many other leading scientists, including Darwin, with whom he continued discussing the theory of evolution. Both men developed a huge respect for each other. Wallace would dedicate one of his books to Darwin as a token of his friendship and admiration. Darwin would tell Wallace that he was far too humble, commenting, "He rates me much too highly and himself much too lowly."

However, despite his outstanding achievements, Wallace struggled to secure regular employment and carried on selling his specimens to earn a living. He went on to write twenty-two books and hundreds of scientific articles on a variety of topics and gave many public lectures, both in Britain and North America, to supplement his income.

> "I sincerely wish you all some of the delight in the mere contemplation of nature's mysteries and beauties which I have enjoyed, and still enjoy."

In April 1866, Wallace married Annie Mitten, who shared his love of nature. They had three children: Herbert, Violet, and William. Together they enjoyed taking long walks in the country and growing unusual and interesting plants in their garden. The family moved many times from one house to another, trying to escape the rapid spread of towns and cities and seeking the tranquility of the countryside.

Wallace lived to the great age of ninety. During his life, he received many prestigious awards for his work as a prolific collector and an extraordinary scientist. He was a humble and self-taught man, with a boundless curiosity and enthusiasm to understand the world around him. Not only did he gather a collection of hundreds of thousands of specimens, many new to European science, but together with Darwin, he brought us a new understanding of how life evolved on Earth. Against incredible odds, the little boy who, years earlier, caught fish in a saucepan in the River Usk had become one of the most famous scientists in the world.

Letter from Charles Darwin to Alfred R. Wallace, April 20, 1870

> "Very few things in my life have been more satisfactory to me — that we have never felt any jealousy towards each other, though in one sense rivals."

Alfred R Wallace Age 25

"Had my father been a moderately rich man and had supplied me with a good wardrobe and ample pocket money; had my brother obtained a partnership in some firm in a populous town or city, or had established himself in his profession, I might never have turned to nature as the solace and enjoyment of my solitary hours, my whole life would have been differently shaped, and though I should, no doubt, have given some attention to science, it seems very unlikely that I should have ever undertaken what at that time seemed rather a wild scheme, a journey to the almost unknown forests of the Amazon in order to observe nature and make a living by collecting."

Alfred R. Wallace

Age 90

EXPLORATION AND EVOLUTION GLOSSARY

adaptation: the process of change by which a plant or animal becomes better suited to its environment

artificial horizon: an instrument used alongside a sextant to provide an artificial horizon when the real horizon is not visible

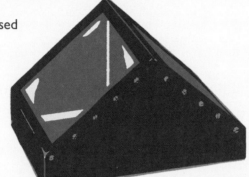

barometer: an instrument to measure atmospheric pressure and predict changes in the weather

botanist: a person who studies plants

canopy: the layer of the forest where the trees spread out their leafy branches like a wide umbrella

characteristic: a feature, trait, or quality of a living thing

classification: a system for grouping living things that share similar features to identify and compare them

common ancestor: the most recent shared ancestor of two or more species

community: a group of populations that live and interact with one another in an area

distribution: the geographical area over which a population is spread

entomology: the study of insects

evidence: information gained from observations and investigations that is used to create logical explanations and answer questions

evolution by natural selection: the theory proposed by Wallace and Darwin to explain the process by which living things gradually change or evolve over time. Individuals with characteristics most suited to their environment are more likely to survive and reproduce than others, passing these useful features on to their young. Over many generations, these gradual changes result in the formation of a new species.

extinct: an animal or plant species that has died out

fauna: word used by scientists to refer to the wild animals of a region

flora: word used by scientists to refer to the wild plants of a region

fossil: the remains or traces of a plant or an animal that died a long time ago, which have been preserved in rock

geologist: a person who studies the origin and structure of Earth and its rock formations

inheritance: characteristics passed down from parents to their offspring

malaria: a disease that is passed on to humans by mosquitoes in hot, tropical countries

naturalist: a person who studies the natural world

organism: a living thing

origin of species: how species come into existence

plantation: a very large farm or estate where crops such as cotton, coffee, and sugarcane are grown on a vast scale, often in a tropical country

population: all of the members of a single species living in the same area at the same time

prismatic compass: a compass that is fitted with an eye slit and a small prism, allowing the user to read the direction of a certain point in relation to magnetic north

Royal Geographical Society: a British institution founded in 1830 to promote the advancement of geography (originally called the Geographical Society of London)

sextant: an instrument used to measure the angle of the sun and stars to determine one's position on Earth's surface

species: a group of plants or animals that share the same main characteristics and can breed with one another

specimen: an organism or fossil that has been collected and preserved for display or scientific research

surveyor: a person who determines the boundaries and elevations of land or built structures

theory: a proposed explanation as to why or how things happen in the natural world

variation: the differences in characteristics between members of a species, such as size or color

yellow fever: a disease spread through mosquito bites in hot, tropical countries

OTHER RESOURCES

Bernardi, Giacomo. "Tracing Alfred Russel Wallace's Footsteps Through the Jungles of Borneo." *Smithsonian*, August 15, 2018. https://www.smithsonianmag.com/science-nature/tracing-alfred-russel-wallaces-footsteps-through-jungles-borneo-180970010/

Cumming, Vivien. "The Other Person That Discovered Evolution, Besides Darwin." BBC, November 7, 2016. http://www.bbc.com/earth/story/20161104-the-other-person-that-discovered-evolution-besides-darwin

Markgraf, Bert. "Alfred Russel Wallace: Biography, Theory of Evolution & Facts." Sciencing, June 4, 2019. https://sciencing.com/alfred-russel-wallace-biography-theory-of-evolution-facts-13719062.html

McNish, James. "Who was Alfred Russel Wallace?" The Natural History Museum, January 8, 2018. https://www.nhm.ac.uk/discover/who-was-alfred-russel-wallace.html

National Geographic Society. "Alfred Wallace." National Geographic Education Resource Library, August 23, 2019. https://www.nationalgeographic.org/encyclopedia/alfred-wallace/

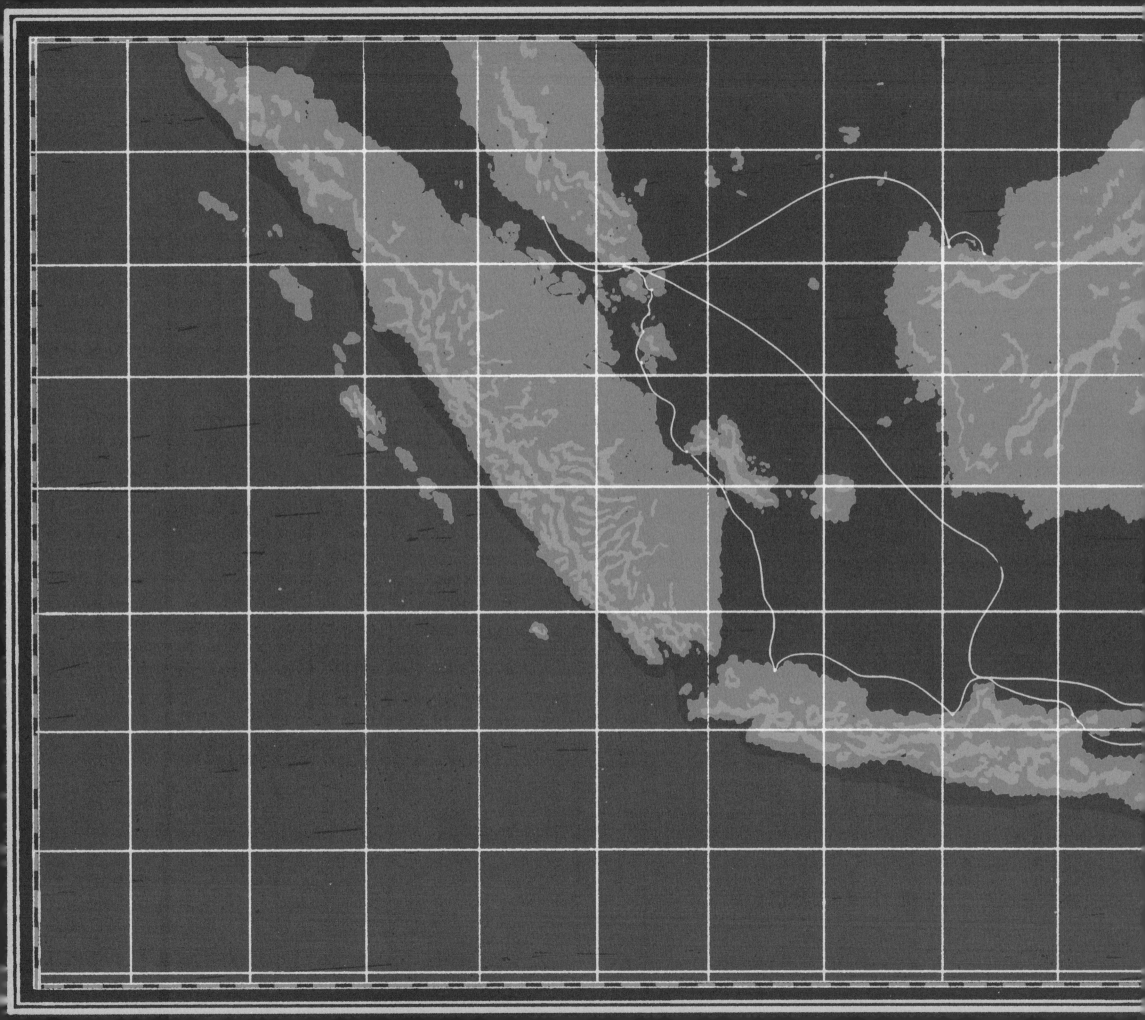